IMAGES
of America

CHATTANOOGA RADIO
AND TELEVISION

Norman Thomas (left) and Earl Winger, childhood friends from Ohio, shared a love of radio, and in 1925, they pooled their resources to put Chattanooga's first radio station, WDOD, on the air. For the first four years, the low-power station at 1280 on the AM dial broadcasted only three nights a week from 7:30 p.m. until 9:00 p.m., carrying symphony and opera selections. By 1929, WDOD expanded, and it would soon broadcast live performances featuring local performers. Thomas and Winger would stay active in Chattanooga broadcasting for almost 40 years. (Both, Earl Freudenberg.)

On the Cover: From 1958 to 1978, school groups and scout troops flocked to the WTVC studio each afternoon to win prizes from Bob Brandy, Chattanooga's favorite singing cowboy. (WTVC.)

IMAGES
of America

CHATTANOOGA RADIO
AND TELEVISION

David Carroll

ARCADIA
PUBLISHING

Published by Arcadia Publishing
Charleston, South Carolina

Printed in the United States of America

Library of Congress Control Number: 2010930147

For all general information, please contact Arcadia Publishing:
Telephone 843-853-2070
Fax 843-853-0044
E-mail sales@arcadiapublishing.com
For customer service and orders:
Toll-Free 1-888-313-2665

Visit us on the Internet at www.arcadiapublishing.com

*To my parents, Hoyt and Ruth, who allowed me to pursue
my love of broadcasting, and to my wife, Cindy, and
sons, Chris and Vince, who encourage me each day*

CONTENTS

ACKNOWLEDGMENTS

As this project began to take shape, I figured I would need a little help—talk about an understatement! Without the help of my friends, this book would have been a pamphlet. I'll begin with Cleveland Wheeler, who started as a local teen deejay to eventually become a nationally recognized radio programmer. He recommended posting some vintage photographs on the Internet, and soon people began suggesting a book. The most persuasive was my colleague Bill Markham, who brought me the Arcadia books about Birmingham and Nashville broadcasting and said, "David, you need to do this for Chattanooga."

So the collecting process began, with contributions from Bob Boyer, Carmen Davis, Kathryn Dunn, Johnny Eagle, Doris Ellis, Tommy Eason, Earl Freudenberg, Danny Howard, Tracy Knauss, Morty Lloyd, Betty Mac, Bob Nolan, Dorris Prevou, Mike Scudder, Booker Scruggs, Brian Stewart, Jeff Styles, Tom Tolar, and Kevin West.

Thanks also to those who shared their memories, helping with the details from Chattanooga's rich 85-year-old broadcast history. Wayne Abercrombie, Margaret Feagans Brown, Harmon Jolley, Marilyn Lloyd, and Bill McAfee were generous with their time.

I was fortunate enough to have interviewed some great local broadcasters who are no longer with us. Pioneers like Bob Brandy, Harve Bradley, John Gray, Bill Gribben, Roy Morris, and Emroy Williamson are all a part of this book, thanks to entertaining conversations recorded years earlier.

Arcadia author Lee Dorman and publisher Maggie Bullwinkel gave me great advice and direction. Chattanooga journalists Barry Courter and John Wilson publicized the project, helping me obtain photographs from people near and far.

It says a lot about Chattanooga's broadcasting brotherhood that every television and radio station put aside competitive differences to help me compile the first photographic history of the city's broadcast heritage. To all those who opened their doors and greeted me with a smile in this most enjoyable project, I thank you.

INTRODUCTION

When I began collecting and sharing old Chattanooga broadcasting photographs, the response was heartwarming. People felt like they were seeing old friends. It is hard for young people to understand within in the context of today's fragmented 500-channel TV universe that once there were only a handful of TV and radio personalities, and those on the local scene were major celebrities.

Chuck Simpson's fans could fill a downtown theater in the 1930s. A few years later, Luther Masingill would begin an unprecedented seven-decade run as the king of Chattanooga morning radio. Gus Chamberlain's daily broadcasts of Lookouts baseball games were considered must-listen radio for a generation of Southern League fans. When the local television signals allowed Chattanoogans to forego the snowy images from Nashville and Atlanta, they were mesmerized by the deep voice and shaved head of Mort Lloyd. Kids learned to tell time from Miss Marcia in the morning and enjoyed the cowboy tunes of the Bob Brandy trio in the afternoon. On weekends, Harry Thornton hosted sellout audiences for live wrestling in studios and Tennessee Valley auditoriums. Each year, Roy Morris would stay on the air for 21 consecutive hours, raising money for the March of Dimes. Nobody could relate to the 1960s rock and roll rebels like deejay Tommy Jett. It's no wonder that they, along with other baby boomer–era icons, are as familiar to Chattanoogans as Walter Cronkite, Johnny Carson, and Jerry Lewis.

Entire books could be written about some of these multitalented, generous personalities. In fact, this book contains a whole chapter about Luther Masingill, whose accomplishments have earned national acclaim. But in order to present a thorough representation of the people who have entertained and informed Chattanoogans since 1925, I set out to include photographs and details of some of the lesser-known figures as well. Some were among the first to staff an announcing booth, deliver breaking news, or do live commercials, while others toiled behind the scenes. All certainly deserve to be remembered and recognized.

I have experienced some memorable moments while producing this book. Some were heartbreaking. There was the son of a long-ago anchorman, expressing gratitude that his father would be included. "He thinks no one remembers him," he lamented. The widow of a popular broadcaster shed tears while going through picture albums she hadn't opened in 20 years. A son of a recently deceased radio deejay thanked me for discovering previously unseen photographs of his dad. He had only one image of his father—the one from his obituary.

There were joyous moments also. I explained to many who are still in the public eye that this was to be a history book and older photographs were preferred. So reluctantly, but with good humor, they would yield a photograph or two from a few decades ago, when hairstyles and fashions were decidedly different, and in their eyes, not particularly flattering. On other occasions, what was intended to be a quick visit to pick up a single picture turned into an afternoon of reminiscing and laughing. More than once, someone would delay my departure by saying, "I just remembered something. There is one more box you might be interested in." And at the bottom of that box would be a treasured photograph, perfect for this book.

As a first-time author, the process was overwhelming at times, especially when balancing the deadlines of a book with responsibilities of home and work. One frustrating day, my friend Tracy Knauss said, "David, you were born to write this book." Perhaps he was right, as my earliest memories include radio and television. My older sisters, Brenda and Elaine, played their radios and records constantly, cheering me on when I would grab something resembling a microphone and pretend I was broadcasting. When I was a teen, my parents were puzzled by my fixation on becoming a disc jockey, but they let me live out my dreams.

I'm grateful I got into this fascinating business. I have been lucky to get to know many of my childhood heroes, like Luther, Miss Marcia, and the late Roy Morris. Sadly, others like Mort Lloyd and Harry Thornton passed away before I could meet them; however, seeing their photographs and reviewing their achievements has rekindled good memories for me. Hopefully, this book will ensure their legacies live on.

After reviewing 85 years of broadcasting in Chattanooga, some realities become evident. In many ways, local radio and television has come full circle. In the beginning, most new stations depended upon network programming to help fill their schedule. The stations had not yet established any personalities of their own, and network shows were convenient and accessible. Eventually, in what many of us now consider the golden age of local broadcasting, stations relied less on network programming, providing local shows ranging from news to entertainment each day. Radio stations were locally staffed with announcers around the clock, seven days a week. As for television, in addition to local news, weather and sports, stations offered programs aimed at housewives, kids, music lovers, and wrestling fans. There were even controversial talk shows, long before the era of Limbaugh and O'Reilly.

At the dawn of the 21st century, many of those local elements had disappeared from the airwaves. As consolidators took control of once locally owned broadcast outlets, syndicated programming began to infiltrate local radio. Much like radio's early days, it is somewhat rare to hear a live, local voice coming from your speakers. Increasingly the deejay you hear playing your favorite song is actually a recorded voice from outside Chattanooga. Though local television allots more time for news than ever before, only recently have there been signs of reducing the amount of syndicated game, talk, and courtroom shows in favor of the live programs that helped make local stations successful in the first place.

It is my hope that the publication of this book might encourage the next generation of broadcasters to understand that Chattanooga has a personality all its own. There is no substitute for good local broadcasting, and this book exists to celebrate those who did it so well for so long.

One

SIGNING ON: THE DYNAMO OF DIXIE

From its first broadcast on April 13, 1925, WDOD (originally AM 1280, later 1310) had a lock on local listeners. In fact, the Wonderful Dynamo of Dixie had a monopoly on the local airwaves for 11 years. Founders Norman Thomas and Earl Winger manufactured crystal radio sets as a hobby and opened the Chattanooga Radio Company to sell those products. They realized Chattanoogans would be more likely to buy them if they could receive a local signal. Some could pick up Atlanta's WSB or other distant stations, but reception was fuzzy at best.

WAPO (1420) arrived in 1936, and four years later, Lookouts baseball team owner Joe Engel entered the fray, adding WDEF (1370) to the radio dial. By then, NBC had two radio networks, and CBS had another, each with established commentators and entertainers. Gradually the three Chattanooga stations would supplement network programming with local morning shows, midday man-on-the-street interviews, afternoon broadcasts featuring live and recorded music, and late-night programs, long before Johnny Carson and Jay Leno tucked everyone in. In 1946, WAGC (1450) signed on, affiliating with the Mutual Broadcasting System. In 1948, WDXB (1490) made its debut, specializing in the hillbilly music of the era, followed three years later by WMFS (1260), which appealed to the rhythm and blues audience. Religious and educational programs also made up a substantial part of each station's schedule.

The late 1940s also saw the city's first attempt to tap into the new FM band. The International Ladies Garment Union put WVUN-FM (98.1) on the air. WDOD added an FM signal (93.7) in 1950. Unfortunately, there weren't enough FM-equipped radios to make a dent in the marketplace, and the stations soon went away. AM stations would continue to dominate the airwaves for the next 20 years, delaying any initiative for Chattanooga broadcasters to invest in the FM band until the 1960s.

Chattanooga radio's first 30 years showcased several young entertainers who would gain national fame and some creative announcers who were learning how to get the city's attention.

WDOD expanded to a 13-hour daily schedule in 1929. The station hosted live *Noonday Frolic* and *Saturday Night Barn Dance* programs at the WDOD Playhouse, recruiting several country western entertainers. Among them was Georgy Goebel, who had performed as a teen on the *WLS National Barn Dance* in his hometown of Chicago. Under the name George Gobel, he later had his own NBC television show and became a regular on *Hollywood Squares*. (Earl Freudenberg.)

East Tennessee native Archie Campbell was barely in his 20s when he began playing the character of Grandpappy. He told jokes and played music, delighting live WDOD audiences daily from 1937 until 1941. Occasionally CBS picked up the shows to run on the network. Campbell later joined the *Grand Ole Opry* and television's *Hee Haw*, sometimes returning to Chattanooga to film commercials for local sponsors. (Earl Freudenberg.)

"Good morning, Breakfast Clubbers, let's twirl some platters!" In the 1930s, the biggest celebrity in Chattanooga was Chuck Simpson. WDOD's ace announcer made news with his comings and goings, taking a job in Miami and then returning after four months to a hero's welcome. Even his speeding tickets were reported in the newspaper. After his morning duties, he conducted man-on-the-street interviews at lunchtime. (Earl Freudenberg.)

Chuck Simpson hosted the *WDOD Barn Dance* each weekend in addition to his weekday morning show. The WDOD Playhouse was located in the Capitol Theater on Market Street. Simpson introduced an ever-changing roster of entertainers, most of whom performed country western music or comedy. He began at WDOD in 1935, and when competitor WAPO signed on in 1936, he touted WDOD as "the preferred station in the territory." (Earl Freudenberg.)

Post–World War II Chattanoogans remember John Totten as the guitar-strumming owner of Totten's Furniture, but at the WDOD Playhouse, he was "Slim" Totten, yodeling, singing, and picking. He later used his musical skills to great success in radio and television commercials, promoting his store. Totten also hosted a Sunday morning gospel music show on WFLI radio for more than 30 years. (Millie Totten.)

By 1941, WDOD had furnished a luxurious studio in the Hamilton National Bank Building at the intersection of Seventh and Market Streets. The station employed a five-person engineering team. Many of them interacted with the announcers on their shows. From left to right are Ernie Feagans, Charles Stokeley, Buel Anthony, Henry McKinney, and Merrill Parker. (Earl Freudenberg.)

In the 1940s, WDOD was still owned and operated by the station's founders. Feeling the heat of increasing competition, the station beefed up its announcing staff, played more recorded music, and emphasized news and sports. Announcers were called upon to take on double duties, hosting music shows while also handling news and sports assignments. From left to right are Bob Bosworth, unidentified, John Gray, and Chuck Simpson. (Earl Freudenberg.)

A Purple Heart recipient in World War II, Roy Morris began a 55-year Chattanooga broadcast career in 1950. "Cousin Roy" worked at WAGC, WDOD, and WAPO before moving to television in 1956. *Laff 'n Live* was the title of Roy's show, and he delivered plenty of humor and audience participation. *RM in the PM* was a late-night tradition for thousands of Chattanoogans. (Pat Treppard.)

John Gray's booming baritone graced the city's airwaves for 40 years. Known for being a sharp-dressed man, John was chief announcer at WDOD from 1946 to 1953, hosting a music show called *Gray's Array*. The Iowa native had served in World War II and worked in radio in Chicago and Atlanta before coming to WDOD. He and his wife, Amanda, operated an antiques business, also called Gray's Array. (Earl Freudenberg.)

William "Peco" Gleason was the original program director for WDEF radio in 1940 as it began competing with the more established WDOD and WAPO stations. He specialized in sports, calling Lookouts games for Joe Engel, who owned both the team and the station. He also conducted Chattanooga's only audience participation show as Colonel Wise, live from the Volunteer Theater. (Earl Freudenberg.)

In 1929, former Washington Senators player and scout Joe Engel was sent to Chattanooga to run the Lookouts minor-league franchise, eventually buying the team. He quickly became known as "baseball's one-man circus" for his flashy promotions. In 1940, he added radio to his repertoire, introducing WDEF to the airwaves. Early on, many folks just referred to it as "Joe Engel's station." (Earl Freudenberg.)

Joe Engel's Chattanooga Lookouts set high standards in Southern League play, and fans were treated to talented, long-term announcers. Gus Chamberlain was at the microphone from 1953 to 1965, following in the footsteps of Arch McDonald, who called the games in the 1930s, and Tom Nobles, the announcer during the 1940s. Chamberlain also served as a WTVC-TV sportscaster in the early 1960s. (Chattanooga History Center.)

Although WDXB arrived in 1948 with programming aimed at white listeners, "Pappy" Ted Bryant soon convinced the station to let him play music for the city's black community at night. His show ran for seven years. The city's first black announcer also owned a successful printing company for 60 years. Shortly after his death in 2004, his groundbreaking radio career inspired the play *Ebony Airwaves*. (African-American Museum.)

Charles "Bugs" Scruggs was a protégé of Ted Bryant, filling in for the WDXB deejay on occasion. The station soon hired him for the early-morning slot. When WMFS, Chattanooga's first all–rhythm and blues station, signed on in 1951, Scruggs was chief announcer and morning host. Scruggs later enjoyed a long career in radio management in Memphis, where he currently hosts a children's television program. (Booker T. Scruggs.)

In early local radio, most live broadcasts could only be conducted in-studio or at nearby locations after elaborate setups. In breaking news situations, enterprising reporters would take a tape recorder to the site, describe the scene, and then rush the tape back to the station. In this dramatic 1951 photograph, WDOD announcer Bill Davies reports on a major fire at the Sears Roebuck store on Market Street. (Earl Freudenberg.)

From 1948 to 1956, many Chattanoogans enjoyed the country western flavor of Charlie "Peanut" Faircloth with their morning coffee. A musician himself, Faircloth hosted square dances and acted as emcee when big stars performed at Memorial Auditorium. After his stint at WAPO, Faircloth worked in radio in Dalton, Georgia, and back in Chattanooga at WDOD while continuing to perform with his band. (Frances Faircloth.)

In 1952, Evelyn Cato entered a contest for a chance to host a radio program on WMFS. She won and began hosting a recipe program called *Listen Ladies* as well as a jazz show, becoming the city's first black female radio personality. After leaving the station in 1958, she served as a secretary for two of Chattanooga's top political figures, C. B. Robinson and John P. Franklin. (Doris Parham.)

Ernie Feagans began as an engineer at WDOD in the early 1940s and would eventually become one of Chattanooga's most enduring personalities during the next half-century. His nighttime WDEF radio show *Feagans Follies* was a top-rated attraction from 1949 to 1954. He later hosted the popular college football scoreboard show *Hold That Line* and reported from major golf tournaments on the PGA Tour. (Margaret Brown.)

Two

THE MAN WITH
SUNSHINE IN HIS VOICE

William Luther Masingill was a frequent visitor to the principal's office at Hardy Junior High School, receiving regular paddlings for talking and fooling around in class. By the time he moved up to Central High School, he was still talkative but learning to use his gift of gab more constructively. Young Luther was called upon to host school talent shows and deliver the morning announcements. Like many teens, Luther sought part-time work and was hired at a service station. He says his love of broadcasting may have begun when he would speak over the store intercom, saying, "Mr. Penney, a gentleman wants to buy some tires." The other workers told him, "You sure sound good, Luther!"

As fate would have it, the 18-year-old attendant was washing the windshield for Joe Engel one December day in 1940. He had heard that Engel would soon open a new radio station and asked his famous customer if he could help answer phones for the announcers. Engel told him to apply at the station, but when Luther arrived, he was given some scripts and told to audition for an announcing job. Despite mispronouncing the word "salon" as "saloon," he was hired on the spot, and was soon hosting the *Jitterbug Jamboree*. A few years later, a newspaper writer coined a phrase that said it all: "He's the radio announcer with sunshine in his voice."

Dominating the local radio scene like no other broadcaster in any market, his recipe for success is simple—"It's more than just playing music. It's what you do between the songs that matters, and I like to help people." Luther has reunited thousands of lost dogs with their owners, asked his listeners to help families whose homes were destroyed by fire, and has promoted an impressive amount of charitable events, perhaps more than anyone else in the world.

Seven decades later, when asked if he had any idea he would still be working in the same time slot at the same station longer than anyone in the history of broadcasting, he replied without hesitation, "You know, I believe so. I just love what I do! I always have."

Chattanooga's Central High School boasts many famous graduates, but the best known is this fair-haired member of the class of 1941. Already a campus celebrity for his outgoing personality, Luther Masingill (center) would frequently interview fellow students on camera—in fact, the filmstrip still exists today. The fair hair wouldn't last long, but the broadcasting career would surpass all others. (Luther Masingill.)

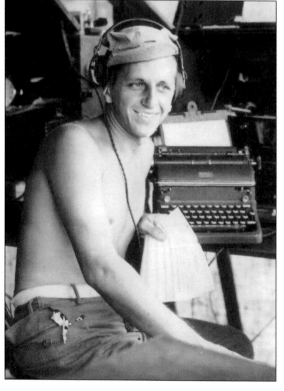

Like many young Americans in the early 1940s, Luther's stateside career was put on hold for a stint in World War II. As he has often joked, "I fought and I fought . . . but they made me go anyway!" Luther served from 1942 until 1945 in the Army 135th Airborne Signal Corps. This photograph shows him in New Guinea. Even today, he uses a typewriter much like the one seen here. (Luther Masingill.)

WDEF held Luther's job for him when he returned from the service, and the one-time cub announcer blossomed into the city's most popular personality. He soon realized he had taken the right career path, choosing radio over his childhood dream of being a train conductor and "seeing the countryside from a caboose." (WDEF.)

Luther's popularity exploded in the 1940s, prompting WDEF to expand his duties. After his morning *Sundial* program, Luther returned in the afternoon to host the *1400 Club* (the station's new AM frequency), later renamed *Loafin' with Luther*. The afternoon music was more upbeat, in an effort to appeal to the teenage after school audience. WDEF would return to 1370 on the dial in 1946, and it remains today. (WDEF.)

During the 1950s, Luther, above with a tour group, received many job offers. Southern entertainers like Ernie Ford and Jimmy Dean were scoring big nationally, so radio executives in Philadelphia, Chicago, and Boston recruited Luther. They were impressed by his ratings, with 62 percent of listeners tuned in each morning. Why didn't he make a move? "I love Chattanooga," he said. "I love the people and the scenery." (WDEF.)

Seen here with Mayor Rudy Olgiati (left) and Ray Solomon, Luther had won just about every local Man of the Year award by the 1950s. Later in his career, he also won the Marconi Award from the National Association of Broadcasters and has a Communication Professorship in his name at the University of Tennessee at Chattanooga. (Luther Masingill.)

Luther was the perfect employee to promotion-minded Joe Engel. He was willing to do just about anything to get attention for his program and for WDEF. On-air feuds, stunts, and contests were commonplace, but Luther was becoming best known for finding lost pets. He was also the radio host parents and students most depended on for school closings information. (WDEF.)

When WDEF added a television studio to its cramped quarters on the fourth floor of the Volunteer Building, station president Carter Parham (right) recruited Luther as chief announcer and afternoon variety show host. Parham's top priority was community service, and Luther seemed to be involved with every charity and civic group in the city. (WDEF.)

FOR LUTHER

Luther jokingly conducted a mayoral campaign in 1951, making outlandish promises in an effort to recruit supporters. As listeners suggested planks for his platform, Luther would readily agree. He pledged to abolish parking meters, reduce the price of haircuts for balding men, build a 12-lane bridge across the Tennessee River, and eliminate railroad crossings in the downtown area. There was only one problem: an alarming number of citizens were planning to actually vote for their favorite morning announcer, much to the dismay of the real candidates. Two days before the election, Luther took out a full-page newspaper ad, making sure everyone was in on the joke. Luther's political escapade inspired a cartoon by *Chattanooga Times* artist W. C. King and was reported nationally by *Time* and *Readers Digest.* (Both, WDEF.)

Luther was involved in constant promotions and giveaways, which were met with enough entries to fill up a truck bed. One memorable contest offered his services for a day to a lucky listener. The housewife who won "worked me to death," he said. "I had to wash her windows, cut her grass, and polish her car." (WDEF.)

When Luther proposed to church secretary Mary Varnell in 1957, it was big news. The two had met at Avondale Baptist Church, and word soon spread that Chattanooga's most famous bachelor would finally tie the knot at age 35. Their visit to the courthouse to apply for a marriage license was captured on film by a WDEF photographer. (WDEF.)

Luther's afternoon television variety show was a showcase for local entertainers in the 1950s. From left to right are Luther, Barbara Molloy, Barbara Delaney, and Bill Wilson. Molloy and Delaney were singers who appeared regularly on local television, and Wilson was a musician who also directed for WDEF-TV. (Barbara Molloy.)

Luther understood the power of personality, sharing details of his life with his listeners. He talked about his wife, Mary; children, Jeff and Joanie; and his dog, Winston. When he rebuilt his beloved 1926 Ford Model T Coupe, he kept everyone informed on the status of his project. His folksy style is a major reason some of his advertisers have stayed with him for more than 50 years. (WDEF.).

By the early 1960s, Luther was married with two children and had cut back on his stunts and personal appearances. His *Sundial* program was a perennial ratings winner, pulling in 50 percent of the audience against six competing stations. Despite the onslaught of rock and roll, Luther remained loyal to his pop music standards, cheerfully urging his listeners to "wake up and hit that cold linoleum!" (WDEF.)

This playful photograph from *Chattanooga Life and Leisure* illustrates the battle for morning listeners that had developed by the late 1980s. There were now 14 stations, with half of them on the FM dial, and some were chipping away at Luther's still formidable ratings. Among the top challengers was WSKZ (KZ-106), represented in this tug-of-war by *Morning Zoo* personality "Dancin' Dorothy" Oliver. (Tracy Knauss.)

The late Buddy Houts was an engineer for WDEF who provided comic relief for Luther. After leaving to become an editor for the *Chattanooga News–Free Press*, Houts would call Luther each day with humorous observations. Often he would tell Luther that he had something important to say, only to disconnect the call, leaving Luther hanging. Luther calls Buddy "the funniest man I ever knew." (Luther Masingill.)

Although Luther's seven-decade run on the radio receives most of the attention, he has also appeared daily on Channel 12 since the station signed on. Perhaps among the most amazing facts of his long career is that in a 54-year period, spanning from 1947 to 2001, Luther never once called in sick. He's seen here with noon newscast anchors Patrick Core and Linda Edwards. (WDEF.)

Three

Lights, Camera, Action!

In 1954, the few Chattanoogans who owned televisions had to be content with snowy images from Atlanta or Nashville stations. Beginning in the late 1940s, officials with four local radio stations assembled proposals in an effort to convince the Federal Communications Commission to grant them Chattanooga's first television license. All submitted programming schedules, staff rosters, and facility plans. Eventually some of the applicants pulled out of the race and joined the WDEF group. In January 1954, WDEF got the nod, and on April 25, 1954, Channel 12 premiered to much fanfare.

Being the first station provided WDEF the opportunity to choose the top programs from both NBC, its primary network, and CBS. The station also made good use of its radio personalities, offering local viewers some familiar names and voices.

After losing out in the competition for Channel 12, WDOD made another run at the television business but would again fall short. Another radio competitor, Ramon G. Patterson of WAPO, was granted the license for Chattanooga's second television station and named it after his initials, WRGP. On May 6, 1956, the new Channel 3 premiered with NBC programming, with Channel 12 opting to carry a full CBS schedule. (WRGP would change hands in 1963, and new owner Rust-Craft Broadcasting would rename the station WRCB.)

In 1957, WDOD cofounder Earl Winger made one last stab at television. He purchased the failing WROM-TV in Rome, Georgia, and transferred the license to Chattanooga, where there was more advertising revenue. He then sold the station to the Martin Theaters chain, which introduced WTVC Channel 9 on February 11, 1958. At first, the new ABC affiliate broadcast on a limited basis, starting its schedule each day at 3:00 p.m. with *American Bandstand*.

WTCI Channel 45 entered the market in 1970, providing public television programming. Two years later, WRIP Channel 61 AM-FM radio station, a Rossville, Georgia, affiliate, signed on, changing to WDSI in 1983 and incorporating with the Fox network in 1986. A sixth full-power station, WFLI Channel 53, began broadcasting in 1987.

This art card was the first image broadcast by Channel 12. It was accompanied by the booming voice of Peyton Brien introducing the station to Chattanooga viewers. The station began broadcasting in cramped quarters in the Volunteer Building, eventually moving into the roomier Glass House restaurant building on South Broad Street in 1958. (WDEF.)

Art cards like the one above and other WDEF station designs were the work of a gifted artist from Nebraska named Emroy Williamson, who was among the station's first employees. "Mr. Willie" created backdrops for children's shows and designed the station's news and variety show sets until his retirement in 1993. (WDEF.)

WDEF's first children's show, *Mr. Chickaroonie and His Friends*, which featured Warren Herring as Mr. Moon and Mildred Gaither as Auntie Mil, aired each afternoon from 1954 to 1958. It was set in an enchanted forest with a gingerbread cottage. The title character, an owl, got his name from a favorite expression of Luther Masingill. In this photograph, Mr. Moon is about to enjoy a delicious Bosco drink. (WDEF.)

In today's crowded television universe, it is hard to imagine a local show drawing hundreds of admiring fans to a personal appearance, but it was commonplace in the 1950s. When the *Mr. Chickaroonie* cast went on the road to the Lake Winnepesaukah amusement park in Rossville, Georgia, it attracted a large audience. (WDEF.)

Barbara Molloy began singing at the age of nine on Rev. T. Perry Brannon's daily WDOD program. When television came along, she became a regular on Channel 12's *Jalopy* variety show and several Channel 3 programs. She's shown here on the *Roy Morris Show* with cameraman Bob Quattlebaum. (Barbara Molloy.)

"Big Jim" Hill was another from WDEF radio's strong stable of personalities who did double duty on the TV side. Luther's frequent radio sidekick became Chattanooga's first TV weatherman. Unlike some radio announcers who feared the arrival of television, WDEF's staff embraced the camera, correctly predicting there would be room for both mediums. (Chattanooga Library Database.)

Also making the switch from WDEF radio to Channel 12, pioneer female broadcaster Drue Smith brought her *Party Line* program to area viewers each weekday at 2:00 p.m. Having worked at WAPO and WDOD in the 1940s, she proved to be a natural for TV with her bright outfits and colorful personality. She spent most of her later career in Nashville reporting for the Tennessee Radio Network. (WDEF.)

Top 10 Dance Party was WDEF's 1957 version of *American Bandstand*, which had yet to be shown in Chattanooga, since there was no ABC affiliate. Cameramen Andy Jonse and Don Turner are shown focusing in on Barbara Delaney and Pete Griffin on the set. (Chattanooga Library Database.)

WRGP Channel 3 hit the airwaves in 1956 with many local shows. After years of backing up Hank Williams and Eddy Arnold, the *Grand Ole Opry's* Willis Brothers settled in Chattanooga with a daily show at noon. From left to right are Chuck Wright with Skeeter, Guy, and Vic Willis. Guy Willis also hosted the afternoon kids show *Wrangler Roundup*, which featured western movies. (Dorris Prevou.)

Like most announcers of television's early days, WRGP's Tom Willette, an original staff member, was called upon to do live commercials, introduce movies, and host a children's show, the *Circle 3 Ranch*. His versatility paid off in later years, when he would host a midday women's show on WDEF and anchor the evening news on WTVC. (Dorris Prevou.)

Channel 3's original facility at 1214 McCallie Avenue featured a small studio that had to accommodate newscasts, commercials, live wrestling, variety shows, and country music stars. The station aired a *Mr. Magoo* cartoon after its *Alex and Elmer* children's show to give the camera crew time to set up the evening news desk. The station moved into a new building near Stringers Ridge in 1968. (Dorris Prevou.)

The WRGP Channel 3 news and on-air staff gathers to show off a shiny new 1959 Triumph convertible inside the studio. From left to right are (seated) Tom Willette and Roy Morris; (standing) John Gray, Mort Lloyd, and Lee Jackson. (Dorris Prevou.)

From 1958 to 1966, WTVC broadcasted from a small studio at its transmitter site on Signal Mountain. Its elevated location caused problems on snowy days, when performers or school groups who were scheduled to appear on live shows could not get to the station. WTVC relocated to the Golden Gateway on West Ninth Street in 1966, followed by a move to the Highway 153 area in 2000. (WTVC.)

Local television's humble beginnings might best be illustrated with this photograph. Unlike the six-figure satellite trucks and mobile news gathering units of the 21st century, WTVC was proud to showcase this decorated vehicle in the early 1960s. The station began broadcasting an evening newscast in 1962. (WTVC.)

Harry Thornton signed on WAGC radio in 1946. By the late 1950s, he was at Channel 3 hosting *Live Wrestling* on Saturday afternoons. In 1966, he moved his wrestling show to Channel 12. Thornton promoted wrestling matches at Chattanooga's Memorial Auditorium as well as in outlying towns. Wrestlers like Jackie Fargo, Tojo Yamamoto, and the villainous Von Brauners were household names in Chattanooga. (Author's collection.)

Jimmy Nabors (shown with singer Barbara Molloy), originally from Sylacauga, Alabama, began working for Channel 3 in 1957. Although his job was editing film, he was also the song-and-dance man on *Holiday for Housewives*. He moved to Los Angeles in 1960, shortened his name to "Jim," and soon became a major star on *The Andy Griffith Show* and *Gomer Pyle USMC*. (Barbara Molloy.)

Just as the children's shows and wrestling matches brought viewers into local studios, so did Channel 12's *Lunch 'n Fun*. Beginning in 1959, the show aired each day at 1:00 p.m., hosted by Jim Garner and Jean Lindsey. Garden clubs and church groups were treated to games, prizes, and of course, lunch. After leaving WDEF, Garner, then known as "Jay," acted on Broadway and guest starred on several television shows. (WDEF.)

In 1959, WRGP's Tom Willette and "Homer the Duck" welcomed Dagmar to Channel 3's studio on the *Tom and Homer Theater* children's show. The voluptuous singer, star on NBC's *Broadway Open House*, was in Chattanooga to appear at a Lookouts baseball game. (Dorris Prevou.)

Four

HEY NOW! WE'RE PLAYING THE HITS

As the 1960s arrived, rock and roll was sweeping the nation. The music of Elvis Presley, Chuck Berry, and other Southern-influenced rockabilly acts provided rich material for the new breed of Chattanooga radio stations. WDOD, which had relied heavily on CBS programming in the 1950s, began cashing in on the hit country sounds out of Nashville. WDXB was first to jump on the top-40 bandwagon, employing popular personalities like Jerry Lingerfelt, Allen Dennis, and Lloyd Payne. It was soon followed by WOGA (formerly WAGC at 1450 AM).

In February 1961, a new challenger arrived on the AM airwaves, shaking up the market like nothing since Luther Masingill's ascendancy 20 years earlier. Radio engineer William Benns got the go-ahead to build a 10,000-watt transmitter for WFLI, 1070 on the dial. Benns installed a youth-oriented format, playing top rock and country hits. Except for mornings, which was still Luther's domain, WFLI, also known as "Jet-Fli," began piling up eye-popping ratings numbers.

Later in the 1960s, FM stations assigned to WDEF and WDOD began making an impact, although they were often simulcasting the stations' AM signals in their infancy. Meanwhile, WNOO (formerly WMFS) maintained its hold on the black listening audience, and dozens of morning announcers attempted to chip away at Luther's massive ratings lead.

Billboard company owner Ted Turner's purchase of WAPO (1150) in 1968 shook up the top-40 market, giving WFLI its first real competition in years. Turner's newly renamed WGOW helped ignite a battle for young listeners that would attract national attention in the 1970s.

By the 1980s, AM's influence was waning, with music fans opting for stronger, clearer FM signals. UT-Chattanooga's WUTC brought National Public Radio to the market, and Chattanooga State signed on an FM station (WCSO, later WAWL) that became a training ground for many of today's top local personalities. Talk radio would become a force, first on the AM dial and eventually on FM. The longtime dominance of WDEF and WFLI was being challenged by powerful 100,000-watt FM stations that livened up the Chattanooga radio marketplace.

Win 1,540 Records From "The Good Guys"

DAVE RANDALL

PAT O'DAY

BILL CARTER

FRED FORREST

BOB RICH

H. C.

PAUL WHITE

BOBBY DARK

HELP WMOC'S "CHICKENMAN" COLLECT FEATHERS!

Shortly after WFLI hit the airwaves, WOGA (1450) became WMOC, joining WDXB in a three-way battle for teens and young adults. The WMOC "Good Guys" were, clockwise from top, Dave Randall, Bill Carter, Bob Rich, Paul White, Bobby Dark, H. C., Fred Forrest, and Pat O'Day. Forrest would go on to become news director at WRCB-TV in the 1970s under his real name, Fred Gault. (Bill Carter.)

The WFLI Light in the Sky guided nighttime listeners to all the happening places in the 1960s. The huge light could be seen from miles in all directions, pointing to advertisers who were hosting WFLI remote broadcasts. If one wanted to meet the deejays and win prizes, one could just follow the light in the sky. (Johnny Eagle.)

Johnny Eagle, one of WFLI's first voices, was program director during the station's 1960s heyday. "Those were the days," he says. "There was no corporate office giving orders. If it was fun, we did it!" Adults controlled the radio in the morning, contributing to WDEF's top ranking; however, WFLI's evening deejays like Tommy Jett and Ed Gale attracted more than half the audience during their shows, designed for after school listening. (Johnny Eagle.)

Just a few weeks after its debut, WFLI welcomed a new nighttime announcer to the staff. At first, this flashy personality called himself Tom Wayne. Owner William Benns suggested something catchier to match the station's flying theme. He said, "How about Jimmy Jett?" The young deejay countered with "Tommy Jett," and he's been the "fastest jet in the air" ever since, greeting listeners with a hearty "hey now!" (Tommy Jett.)

Larry "The Legend" Johnson is shown here interviewing Pres. Lyndon Johnson at the Chattanooga airport in 1964. The WDXB morning man overcame the station's relatively weak signal, attracting advertisers with his energetic promotions. After 15 years on the air, he left Chattanooga for a Milwaukee radio job in 1970, eventually landing at WIND in Chicago. (Ann Terrell.)

WFLI afternoon deejay Nick Smith ("Nic-a-lo, Nic-a-lo, the Lookout Mountaineer") also worked as an engineer at the station from 1962 to 1970. In 1967, he mapped out the station's coverage plan to increase the transmitter's power from 10,000 to 50,000 watts, expanding WFLI's already impressive reach. (Mike Scudder.)

WFLI's Dale Anthony was given the difficult task of going head-to-head with longtime morning ratings champ Luther Masingill from 1961 to 1971. "Doctor Dale" usually won the battle for second place, which was no small feat in the face of increasing competition. He was also a top advertising salesman, later becoming WFLI's station manager. He started his own rock music station (WZDQ-FM 102.3) in 1977. (Judy Graham.)

Twice annually, WFLI presented multi-star concerts known as *Spectaculars* at Memorial Auditorium, featuring the hottest acts on the charts. In this 1968 backstage photograph, deejay Mike Murray (left) is seen with singer Neil Diamond. Mike's afternoon *Murray-Go-Round* show attracted almost three times the listeners of his nearest competitor. He worked at WFLI from 1967 to 1972 and also hosted a nationally syndicated show from the station's Tiftonia studio. (Mike Scudder.)

WFLI's Bill Miller started as a teenage weekend deejay in 1968 and soon worked his way up to program director. The Polk County native grew up listening to WFLI, and even after working in many big cities during his career, he says his heart always belonged to the station. Today he hosts a bluegrass show on WAMU, a public radio station in Washington D.C. that can be heard online. (Judy Graham.)

In the 1960s, WNOO was consistently a top-five Chattanooga radio station, partially in part to a solid lineup of personalities. From left to right are Dave "The Rave" Oliver, "Doctor" Frank Jackson, "Groovin' Grover" Rivers, and "Sweet Daddy" Clarence Scaife. At "1260 on Your Swinging Radio Dial," WNOO played a combination of soul and gospel music selections. (African-American Museum.)

Bill Nash (seated) was a WDOD deejay, University of Chattanooga sports announcer, and sales executive before becoming the radio station's general manager. In 1970, he celebrated WDOD's 45th birthday with several elected officials and sponsors. From left to right are Hamilton County judge Chester Frost, Raymond Proctor, Jack Russell, Jerald Sholl, sheriff H. Q. Evatt, and John Totten. (Earl Freudenberg.)

While he was a junior in high school in 1964, Earl Freudenberg landed a job at WAPO radio. In this 1966 photograph, the young broadcaster mans the booth at WDOD, where he would spend most of his career. He was named the station's program director at age 21. He became widely known in gospel and country music circles, emceeing shows throughout the Tennessee Valley. (Earl Freudenberg.)

WFLI had dominated the teen radio scene for almost a decade when Ted Turner, a McCallie School alumnus, purchased the city's second oldest station. WAPO had fallen upon hard times, and owner Martin Theaters was ready to sell. Turner hired several former WMOC deejays, including Dave Randall (Cleveland Wheeler), and challenged Jet-Fli with WGOW's new top-40 format. (WGOW.)

During WGOW's first year, the station didn't faze WFLI, so Ted Turner recruited a former Atlanta deejay with a proven track record to shake things up. Bob Todd called himself "Chickamauga Charlie" and proceeded to wake up Chattanoogans with a mix of jokes and social commentary. "Chicky-poo" became so popular that WDXB lured him away to create a third top-40 station along with other major market deejays in 1971. (Bob Todd.)

WFLI's staff gathered at Memorial Auditorium in 1970 to take on the Harlem Globetrotters. Warming up are, from left to right, (first row) Neil McWhorter and Frank "Barney Sparks" Dobbs; (second row) Stan T., Rick Shaw, "Young Stanley" Hall, and Gene Lovin; (third row) Ray Anderson, Tommy Jett, and Jimmy Byrd. (Judy Graham.)

Female deejays were scarce in the 1960s, but as the next decade began, women like Jackie Mason were breaking through. Starting at album rock station WSIM 94.3 FM in Red Bank, Jackie later worked at WDXB and WDEF, where she cohosted a two-hour midday show with Luther Masingill. Jackie also forecasted the weather on Channels 3 and 12, adding to her already impressive credentials. (Jackie Mason.)

Billed as "one of America's top country deejays," Jerry Pond, seen above in the WDOD studio in 1975, lived up to the title, earning solid ratings for almost three decades at WDOD, WDXB, and US-101. Jerry knew all the Nashville stars, as they frequently called into his show. (Earl Freudenberg.)

This photograph of Earl Freudenberg was taken at WDOD in 1975. Earl spent more than 30 years at WDOD, where he became Chattanooga's longest-running radio news director. He also hosted *Sound Off*, a popular Sunday morning call-in show on WDOD-FM, and the *Hey Earl* talk show, weekdays on the AM station. In 1981, Earl was named Tennessee Broadcaster of the Year by the Associated Press. In 2005, he switched to WDYN-FM. (Earl Freudenberg.)

In 1962, Illinois native Charles "Chuck" Krause came to work for WDOD, playing country music each evening. The management at WDEF, always on the lookout for the next Luther Masingill, found their man, hiring him in 1964 and renaming him "Jolly Cholly." Hosting the afternoon *Road Show* and handling TV hosting duties as well, Krause remained at WDEF until he died unexpectedly in 1974. He was only 48. (WDEF.)

Bill Poindexter had worked in his hometown of Rossville, Georgia, on 500-watt WRIP radio before hitting the big time at 50,000-watt WFLI in 1973. First using the name "Ron Dean," Poindexter soon switched to his own nickname of "Dexter" and was on the radio each afternoon, becoming program director in 1975. In 1976, he left WFLI to work in record promotion for ABC and RCA. (Author's collection.)

By 1975, WDXB had switched back to adult music, leaving WGOW and WFLI to battle it out for top-40 supremacy. WFLI relied on familiar voices like Jimmy Byrd, a longtime listener favorite. "Fast Jimmy" had started at WFLI in 1969 at the age of 16 while attending Rossville High School. He later moved to Kansas City, Missouri, where he died in 2002 at the age of 49. (Harmon Jolley.)

Capitalizing on the 1975 disco craze, WFLI made a bold move, placing its first black deejay on the air each night at 10:00. The experiment proved successful, as *Quincy Lane's Disco Train*, hosted by Albert C. Fields, was a ratings smash. (African-American Museum.)

50

Milwaukee native Gene Michaels arrived in Chattanooga in 1976 to program top-40 music on WGOW but switched to country radio in 1980. He served up "burnt toast and coffee" on his WDOD-FM morning show and was among the original crew when WUSY (US-101) premiered in 1983. He was working in Lake Charles, Louisiana, at the time of his death in 2009. (Chattanooga Library Database.)

Tyrone Robert "T. R." Gunn was a top Chattanooga radio news director, earning the 1978 Associated Press Tennessee Broadcaster of the Year award for his work at WNOO. If that wasn't enough, Mayfield Ice Cream designed a carton to commemorate his work. (T. R. Gunn.)

wgow 1150

JIM PERKLE

Jim Pirkle was one of Chattanooga's busiest radio personalities from 1974 into the 1990s, excelling in both rock and country formats. He did mostly morning drive shows on WGOW, WFLI, and WDOD. He died on January 5, 1997, at the age of 49. (Author's collection.)

From 1982 to 1986, *Those Guys In The Morning*, Garry Mac (left) and Dale Deason, drew a sizable audience to WGOW's adult-oriented format. Their weekly "Media Madness" segment charted the moves of local media personalities across an imaginary game board. Mac, who grew up in Red Bank, was a longtime deejay and news director at WDXB before joining WGOW, while Deason came from WFOM in Marietta, Georgia, in 1980. (WGOW.)

Unlike the present day, most Chattanooga radio stations had fully-staffed news departments in 1981, much like their television counterparts. From left to right are Channel 9's Mike Lumpkin, WDOD's Mark McNulty, Channel 3's Audrey Ramsey, WDEF Radio's Larry Taylor, an unidentified photographer, Channel 12's Ken Martin, and WGOW's Cindy Carroll, all interviewing Chattanooga Public Works commissioner Paul Clark. (Author's collection.)

Most radio stations fielded basketball or softball teams, and WSKZ (KZ-106) was no exception. The station signed on the air on August 19, 1978, as Chattanooga's first 100,000-watt rock station, becoming the city's top-rated station within a year. The Foul Tips softball team took on the band Alabama in a 1982 charity game, attracting almost 5,000 fans to Engel Stadium. (Author's collection.)

In December 1986, a struggling album rock station transformed as WLMX (Lite Mix 105) and became one of the city's top three stations within a year, despite a relatively weak signal. Bill Burkett (left), who had worked for WGOW in the early 1970s, returned to the market to play straight man to Chattanoogan Parker Smith, who specialized in "did he really say that?" moments every morning. Smith died in 2010 at the age of 48. (Tracy Knauss.)

Patti Sanders played heavy metal music for Rock-105 and followed with a soul/jazz format on WJTT (Jet 94) before settling into the adult contemporary music style of Lite Mix 105 in the 1980s. Most listeners know her as the always-happy midday deejay with a smile in her voice. (Patti Sanders.)

Somewhere between the bulky remote broadcasting equipment of the 1940s and the tiny cell phones of today was the Marti unit. Developed by George Marti in the 1970s, stations like KZ-106 put the radio remote transmitter in a shopping cart and hit the streets to broadcast live from the annual downtown Bessie Smith Strut. From left to right are Al McClure, Mary Dabney, Scott Chase, and Jeff Cooper. (WSKZ.)

"Big Jon" Anthony was one of the original personalities on US-101 (WUSY-FM) when it replaced a short-lived rock station (WQLS) on June 20, 1983. An expert programmer and promoter, he guided the station to the top of the ratings by 1991. US-101 has remained number one ever since, owing a great debt to Anthony. He died suddenly in November 1991 at the age of 45. (Tracy Knauss.)

Known as the "voice of the UTC Mocs" since 1980, Chicago native Jim Reynolds has proven to be a popular and versatile broadcaster, spending almost his entire career with WGOW. In the 1980s, the sportscaster's quick wit was featured on the KZ-106 and WGOW morning shows. Since 1995, "J. R." has been a featured talk show personality on WGOW's *Morning Press* and *Village Idiots* programs. (WGOW.)

The phrase "larger than life" may have been invented for Wally Witkowski. It's not his real name, but his persona has proven to be a hit with talk radio listeners for almost 20 years. Far from politically correct, his verbal jousting with fellow *Village Idiot* Jim Reynolds has attracted a loyal 9:00 a.m. audience on WGOW-FM, the Talk Monster. (Dorothy Oliver.)

In the earlier days of radio, competing stations would engage in a friendly bowling match or a coed softball game. By the 1980s, paintball was catching on, inspiring a messy matchup between US-101 and WGOW. It is hard to tell which team won from this photograph, but both sides seem pleased with the outcome. (Jeff Styles.)

Many radio personalities have conducted talk shows in Chattanooga, but few have enjoyed the success and longevity of WGOW's Jeff Styles. Since the early 1990s, Styles has shared his opinions with listeners, as well as his passion for the environment. (Jeff Styles.)

Jeff Styles came to Chattanooga as a news reporter in the 1980s and moved to Knoxville before returning to host *The Morning Press* and *FRED the Show* ("free radio every day") on WGOW in 1993. For five hours each day, Styles interviews news makers and riles the city's establishment, urging politicians to "just tell the truth!" (WGOW.)

Kevin West arrived in Chattanooga in 1987, just in time to take the reins of WGOW's news department, one of the last surviving radio news departments in the market. The Michigan native delivers hourly newscasts with just a hint of attitude and isn't afraid to express his opinions on the station's talk shows. (WGOW.)

Long before sports talk shows became a national phenomenon, WGOW's *Sport Talk* tapped into a sports-hungry Chattanooga audience, eager to hear inside information on Tennessee Volunteer sports and anything related to college football. Brothers Jerre (left) and Gary "Dr. Basketball" Haskew are in this 1991 photograph debating a hot topic, while producer Scott "Earthquake" McMahen runs the control board. (Scott McMahen.)

"Captain" Bobby Byrd and David Earl Hughes were two of US-101's top deejays in 1995, when the station began an unprecedented run of winning Country Music Association Station of the Year awards. After US-101 won in the medium market category for the seventh straight year, the CMA ruled that stations could not win in consecutive years; thus, US-101 promptly began winning every other year. (WUSY.)

WJTT-FM (94.3) had gone through several formats before finding a winning formula in 1980. The station adopted a soul/jazz format, and many of WNOO's former listeners followed the trend of listening to their favorite music on the FM dial. Keith Landecker has overseen WJTT Power 94 since 1989, and his afternoon *Traffic Jam* keeps them among the city's top stations. (WJTT.)

Larry Ward, also known as the "voice of the Chattanooga Lookouts," has called baseball games for local listeners since 1985, totaling well over 3,000 games. The dean of Southern League baseball announcers is also the longest tenured play-by-play man in the Lookouts' storied history. He even appeared in Alabama's 1994 *Cheap Seats* music video, playing the role of home plate umpire. (Nelle Ward.)

In the fall of 1985, KZ-106 programmers broke away from the traditional single-deejay morning show format to try something radically different. The result was the *Morning Zoo*, and it jump-started the day for thousands of Chattanoogans. Above in this 1989 photograph, clockwise from left, are newscaster Tom Henderson, Jay "The Jammer" Scott, David Hughes, Jim Reynolds, "Dancin' Dorothy" Oliver, and Randy Ross. (Tracy Knauss.)

Chattanooga's undisputed king of soul music is the legendary Bobby Q. Day, whose name is synonymous with WNOO radio. Bobby Q. worked there for almost two decades, starting in 1971. He played everything from jazz to blues. He also hosted a top-rated nightly blues show on WDXB in the late 1980s. (Lido Vizzutti/*Chattanooga Times–Free Press*.)

Jim Copeland began at WFLI in 1974, and then steered WGOW's top-40 format in a more adult direction in the late 1970s. In 1991, he was production director at US-101 when he was tapped to replace the late Big Jon Anthony on the station's morning show. He and Bobby Byrd teamed up to play hot country favorites for four years. In 1995, Copeland moved to WSB Radio in Atlanta. (WUSY.)

In 1995, US-101 discovered a winning afternoon formula, pairing "Big Ol' Hairy" David Earl Hughes (left) with Bill "Dex" Poindexter (right). Dex started out doing traffic reports for Hughes, but their strong chemistry soon made it a two-man show, *Dave and Dex*. They are shown here with singer Blake Shelton. Hughes left US-101 for WSM-FM in Nashville in 2003. He died in 2004 at the age of 48. (US-101.)

Five

WE'VE GOT PERSONALITY!

In the early days of local television, networks left plenty of holes to fill in the daytime schedule. Syndicated talk and court shows were still several years away. Local newscasts filled only a few minutes each day, unlike the multiple hours of today. The founders of Chattanooga television had to be creative.

Many of the early television announcers had come from radio and were accustomed to ad-libbing and audience participation. Some were even performers in their own right. From day one, they turned on the studio lights, welcoming singing groups, Scout troops, women's clubs, guitar pickers, and anyone else who could entertain on live television.

They used homemade puppets and simple props. They cut colorful pictures out of *Life* magazine, flipping through them while music from phonograph records played in the background. They recruited film editors, directors, and receptionists from the station's office to sing, dance, or model the latest fashions. They filled airtime with viewer telephone calls. Taking a cue from morning radio, television hosts started viewers' days in various styles, from congeniality to controversy. Some of this low-budget entertainment caught on and lasted for years. Those personalities that endured through the baby boomer era are among the most familiar names in Chattanooga broadcast history.

When general manager Reeve Owen transferred from Columbus, Georgia, to put WTVC Channel 9 on the air in Chattanooga, he knew he would need a children's show host to compete with similar programs on Channels 3 and 12. Owen recruited his Columbus colleague Bob Brandy, along with his wife, Ingrid, and their trusted Palomino horse Rebel, to ride into town and save the day. (WTVC.)

Bob Brandy, seen with "Phil Ossifer, the Tennessee Talking Mule," sang western songs, played the guitar with his trio, and invited school groups to the WTVC studios. If one were on *Bob Brandy*, one was a star. Bob also showed *Popeye* cartoons and *Three Stooges* comedies, giving the studio crew time bring in his horse Rebel, so the kids could sit in the big western saddle. (WTVC.)

Brandy welcomed guests like *Grand Ole Opry* star Minnie Pearl to his program, which aired at 5:00 on early weekday evenings. He also took his show on the road each weekend, filling school auditoriums from northeast Alabama to western North Carolina. Like the TV show, his live appearances featured stunts, prizes, and snacks. Sit on Rebel, throw a ball into the barrel, and one could win 10 silver dollars. (WTVC.)

In the mid-1970s, Brandy's on-air workload was down to one show a week on Saturday mornings. By 1978, most local children's programs had declined in popularity, so Brandy's remarkable 20-year run was over. After riding off into the sunset, he remained with WTVC as a sales manager and consultant until his death on February 28, 2003, at the age of 72. (WTVC.)

In 1956, Roy Morris followed his WAPO radio boss Ramon Patterson to television when Patterson signed on WRGP Channel 3. The staff announcer made a smooth transition to on-camera work, handling both news and variety assignments. In this 1963 photograph, he is interviewing America's Junior Miss, Diane Sawyer of Kentucky, who would go on to a distinguished broadcast career of her own. (Dorris Prevou.)

The Roy Morris Show aired daily at 9:00 a.m. in the 1960s, and Morris would do anything for a laugh. At the height of Beatlemania in 1964, Morris and his production staffers donned mop top wigs for a lip-synch takeoff of the Fab Four. As the "Four Ticks," they fielded requests for photographs and personal appearances. From left to right are Morris, Irv Prevou, Wayne Abercrombie, and Bill Myhan. (Dorris Prevou.)

During the run of his variety show at Christmastime, Roy Morris brought his children to the set and invited those of his cast and crew as well. The televised holiday party was an annual highlight. (Dorris Prevou.)

A popular annual *Roy Morris Show* promotion gave children a chance to win "A Circus in Your Own Backyard." This lucky 1965 winner got to ride atop an elephant and meet Roy and his cohost Gay Martin. Martin would go on to become the city's first female weather forecaster for WTVC Channel 9 in 1967. (Dorris Prevou.)

At 1:00 p.m. on weekdays in the 1960s, NBC gave its affiliates an hour for local programming, and Channel 3 produced a public affairs show called *Bulletin* in the time slot. Local viewers got a real-life bulletin on November 22, 1963, when the panelists broke the news of the assassination of President Kennedy. From left to right are John Gray, Roy Morris, Joan Barry, and Don Fischer. (Dorris Prevou.)

Betty McCullough, known as "Betty Mac," was a popular personality on Chattanooga television and radio. Her 1960s shows on Channel 12 included *Woman's Whirl* and the *Morning Show*. Always attired in the latest fashions, Betty promoted the arts and performed in plays at the Chattanooga Little Theater. In the 1970s, she continued her career as a host and news reporter for Channel 3. (WDEF.)

In 1961, *Romper Room* was seen in cities across the nation, each with a local hostess. WTVC's "Miss Marcia" Kling was one of the best. After one year of using the magic mirror, and explaining the difference between "do-bees" and "don't-bees," WTVC created its own daily show for Miss Marcia, called *Funtime*. She supplemented her live segments with cartoons including *The Funny Company*, *Clutch Cargo*, and *Bugs Bunny*. (WTVC.)

On *Funtime*, Miss Marcia sang and played an alternate version of *Happy Birthday*, still fondly remembered by thousands of baby-boomers. After all, it was as if she was singing it to them personally. She also assisted her young viewers in tasks like learning to tie their shoes and tell time. When the big hand was on the 12, and the little hand was on the 9, it was nine-o-clock—*Funtime*! (WTVC.)

"FUNTIME" Weekdays 9 A.M.

Miss Marcia taught preschoolers educational basics and good manners for 16 years on *Funtime*, then remained at WTVC as director of community affairs, as well as a cohost of news programs aimed at women and senior citizens. In the 1980s, she created *Nifty Nine*, giving elementary students the opportunity to present news for kids. (WTVC.)

Tom Willette switched from Channel 3 to Channel 12 in the 1960s, putting his ad-libbing talents to good use on the daily *Lunch 'n Fun* show. In addition to the games and contests, the show offered local charities free airtime to promote their fund-raising drives. (WDEF.)

From 1967 to 1973, WRCB Channel 3 staged an ambitious 19-hour telethon each January. From 11:00 p.m. on Saturday to 6:00 p.m. the next day, the *March of Dimes Telerama* was broadcast live from Memorial Auditorium. Camera crews and on-air personalities from competing stations pitched in, and NBC stars like Richard Dawson of *Laugh-In* (standing, right) joined host Roy Morris in making pitches for viewer donations. (WRCB.)

By 1976, the *Jerry Lewis Muscular Dystrophy Telethon* had become a major annual event. Former *Top 10 Dance Party* emcee Neil Miller hosted the local segments for Channel 12 from its studios on South Broad Street. (WDEF.)

Live Wrestling host Harry Thornton surprised his fans in 1969 by starting a WDEF morning program and adopting the new persona of an opinionated talk show host. Originally called *Regional Report*, the show later expanded from 30 to 90 minutes and was renamed the *Morning Show*. Thornton, seen here with cohost Judy Corn, smoked a cigar, argued with callers, and proved there was a lucrative market for local morning television. (WDEF.)

Bob Elmore (left) was a former WRGP advertising salesman who became chief executive officer of the Chattanooga Convention and Visitors Bureau. In 1971, he convinced WTVC to air *Backyard Safari*, a Sunday afternoon series promoting tourism. The show aired for 16 years, attracting guests like Senator Bill Brock. The show's theme song featured the Andrews Sisters singing "Happiness lies / right under your eyes / back in your own backyard." (Bob Elmore.)

WTVC program director Tommy Reynolds was with the station from 1958 to 1976 and is best known for presenting horror movies on Saturday nights at 11:00. The show began as *Science Fiction Theater*, but when Reynolds created the personality of Dr. Shock, the title was changed to *Shock Theater*. Utilizing costume store props and local characters like Nurse Badbody, the show was appointment viewing for teens who stayed up late. (WTVC.)

From its Black Sabbath theme song to Dr. Shock's sign-off warning that "the scream you hear will be your own," WTVC's *Shock Theater* was better known for the local segments than the horror movies. Dan East voiced the puppet Dingbat, shrieking "Flash, flash, flash!" before delivering satirical news stories lampooning area politicians. They knew they had made it if their names were mentioned on *Shock Theater*. (WTVC.)

In 1980, WRCB held auditions for the new weeknight *PM Magazine* series and didn't have to look far to find its female cohost. Weather girl Debbie Baer had been hired away from WDEF a few months earlier and was a perfect fit for the fast-moving show. Bob Austin came to *PM* from Columbia, South Carolina, and his laid-back personality meshed well with Debbie, who put the "perk" in perky. (WRCB.)

PM Magazine was an immediate hit in the 7:30 p.m. time slot for Channel 3, and the show weathered several personnel changes. John Davis replaced Bob Austin in 1982, and Debbie Baer left to host a Nashville television morning show in 1983. The show remained on the air until 1985. (WRCB.)

Harry Thornton considered retirement while fielding job offers from other stations during his 13-year run on the WDEF *Morning Show* but remained loyal to the station until he retired. In July 1982, citing health problems, he told viewers goodbye. He died four months later at the age of 64. Cohost Judy Corn said that despite Thornton's argumentative on-air style, "he was a true Southern gentleman."(WDEF.)

After Judy Corn and Don Welch both left the *Morning Show* in 1983, Channel 12 scrambled to find new hosts for the long-running talk show, which had topped the ratings since 1969. David Carroll of KZ-106 and Channel 12 news reporter Helen Hardin (above with restaurant-chain mascot Wendy and actress Clara Peller) were tapped to take over the show, which eventually ended its run in July 1987. (WDEF.)

The next big trend in Chattanooga morning television was an expansion of news, weather, and traffic. WRCB's *Eyewitness News Today* premiered in 1989 as a 15-minute program, and thanks to its talented team, it soon grew into a two-hour ratings powerhouse. From left to right are producers Karen Hinkle and Courtney Saunders, with anchors Jed Mescon, LaTrice Currie, and Thom Benson. (LaTrice Currie.)

WTVC relied on a hometown personality to carry its morning banner for more than 20 years. Don Welch (left, with meteorologist Bill Race) worked at WDXB radio in the 1960s and then spent seven years at Channel 3 before moving to Channel 9 in 1975. From 1983 to 1987, he hosted *Good Morning Don*, followed by *Good Morning Chattanooga* in the 1990s. (WTVC.)

Six

YOUR TRI-STATE EYEWITNESS ACTION NEWS TEAM

Today Chattanooga viewers are accustomed to seeing five hours of local news a day on some stations. The legendary newscasters of the 1950s had few resources at their disposal, so they didn't take up much time presenting the day's events. Much like the national networks, local stations started with about 15 minutes of news each day, gradually increasing their output.

The city's first news anchors were armed with little more than the daily newspapers and a typewriter. Within a few years, each station hired photographers who were dispatched to cover ground-breakings, accidents, and fires. When news happened shortly before the live broadcast, the freshly processed film would barely make it on the air, often still wet while being threaded through the projector.

In the 1960s, CBS and NBC expanded their evening newscasts to 30 minutes, as did their Chattanooga affiliates. ABC, a relative latecomer to the network news game, was running a few years behind. The same could be said of its affiliate WTVC, which entered the field with a three-man news department in 1962.

The explosion of television news was triggered in part by the 1963 assassination of President John F. Kennedy. The networks had to rely heavily on their Dallas affiliates to help cover the story, which sent a message to local newsrooms around the country. With civil rights marches, the space program, and the Vietnam War providing indelible images, stations began beefing up their staffs. When newscast ratings began to rise, stations managers realized the potential for increased advertising revenue. The channel with the number-one newscast often attracted more viewers in the time periods before and after the news, and a popular anchor team could create station loyalty that could last for decades.

In broadcasting's infancy, it was said that the man with the deepest voice read the news, and no one's was deeper than Mort Lloyd's. When Channel 12 debuted in 1954, the 23-year-old Shelbyville, Tennessee, native had been a radio host and wrestling announcer in Nashville. Lloyd's daily newscast on WDEF-TV was *World News on Report*, airing for five minutes at 6:25 p.m. He also read poetry on WDEF radio. (WDEF.)

Chattanooga's first local television sportscaster was Herschel Nation, who hailed from a prominent local family. Nation, one of the city's best-dressed men, presented a daily 10-minute *Sports Review* on WDEF-TV from 1954 to 1959. (Chattanooga Library Database.)

In response to the 1962 Cuban Missile Crisis, WDEF-TV presented a three-man anchor team on its new *Global News Report*. From left to right are Harve Bradley, Bill Gribben, and Steve Conrad. The three radio veterans provided a solid nucleus for Channel 12's news programming in the early years of the station. (WDEF.)

Election night at the Chattanooga news stations meant all hands on board, with staffers from various departments counting votes and working the phones to get the most accurate results. In this 1962 photograph, Channel 12 anchorman Steve Conrad can be seen in the center of the action. (WDEF.)

New York City–born Conrad Czarnik earned the Bronze Star for courage during World War II. Friends suggested he change his name for radio, and he became Steve Conrad. He worked at WDOD and WAPO from 1951 to 1956 and was Channel 3's first anchorman. He switched to Channel 12 in 1958, resigning in 1967 to run for the Chattanooga City Commission, where he served for eight years. (WDEF.)

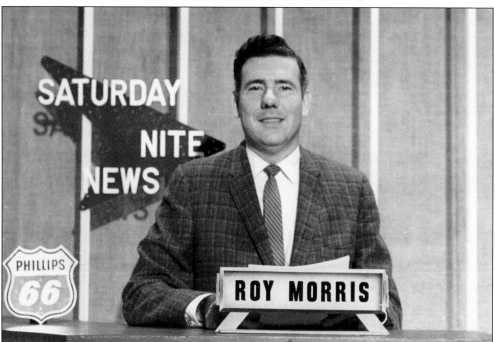

Roy Morris was Mr. Everything at Channel 3, beginning in 1956 and continuing for most of the next 17 years. The *Saturday Nite News* of the early 1960s was actually a brief segment during the *Saturday Nite Movie*. Like most newscasts of the era, it was sponsored by an oil company. (Dorris Prevou.)

Mort Lloyd would shake up the local news scene twice in his career. The first time was in 1958, when he moved from Channel 12 to Channel 3. He and rival Steve Conrad switched places. Channel 3 owner Ramon Patterson told a reporter that Lloyd had more viewers than Conrad, hence the reason he pursued him. The two would compete for nine more years, with Lloyd consistently winning the battle. (Dorris Prevou.)

In the 1960s, live remote broadcasts were major operations requiring stations to move most of their equipment out of the studio to various locations. In 1963, Channel 3 sent Roy Morris and Mort Lloyd to Market Street to televise the annual Armed Forces Day Parade. Morris and Lloyd are in the center of this image interviewing County Judge Chester Frost. (Dorris Prevou.)

Bill Schoocraft did the sports for WDEF-TV's evening news from 1958 to 1968. He was also heard nationally as the Chattanooga correspondent for the NBC radio *Monitor* program, carried locally each weekend by WDEF radio. (WDEF.)

Don Fischer was Bill Schoocraft's longtime rival in sports at Channel 3. The New York native was WDOD radio's sports director from 1956 to 1960, when he began an 11-year run at Channel 3. He also anchored the local news segments during the *Today Show* and was a panelist on the station's midday talk show *Bulletin*. (Dorris Prevou.)

Mort Lloyd's right-hand man during his 20 years in Chattanooga news was photographer Tommy Eason (right), who, it was said, would usually arrive at an accident scene before the police. Eason followed Lloyd from Channel 12 to Channel 3 in 1958 and again to Channel 12 in 1970. In 1979, Eason returned to Channel 3, staying until his retirement in 2009. (Morty Lloyd.)

Mort Lloyd was an accomplished pilot and would often use his plane for news-gathering purposes. In 1958, during a high-profile impeachment trial in Nashville involving a Chattanooga judge, Lloyd flew back and forth to the state capital each day, returning in time to file a report on Channel 3's evening news at 6:30. (Dorris Prevou.)

In 1959, Gil Norwood left Knoxville, Tennessee, at the age of 20 to become an announcer at WTVC Channel 9. At first, the station had only a five-minute daily newscast at 11:15 a.m. Norwood started an evening newscast when he was named news director in 1963. Always short-staffed compared to his more established competitors, Norwood anchored the news until 1970 and served as news director until 1976. (WTVC.)

Bill McAfee joined WTVC in 1963 from a Calhoun, Georgia, radio station, providing a calm, cheery presence on the news until 1975. McAfee handled weather and sports duties in the evening and read the news for the 11:00 p.m. edition. In 1976, he was elected to the Tennessee Legislature, where he served for 24 years. A portion of State Highway 27 north of Chattanooga is named for him. (WTVC.)

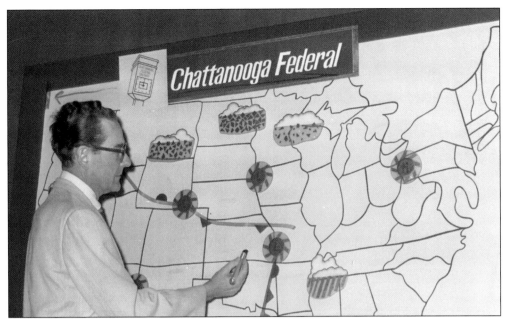

After a long career in Chattanooga radio, John Gray joined Channel 3 in 1957, becoming the city's top-rated weatherman for the next two decades. Long before the computer age, Gray would drive each day to the airport. He gathered data from the meteorologists before compiling his unscientific but reliable weather forecast. Anchorman Mort Lloyd would introduce "Old John" each day with a poem. (Dorris Prevou.)

In contrast to John Gray's no-nonsense style at Channel 3, Harve Bradley brought some levity to the weather on Channel 12 from 1960 until 1967. "The South's most unpredictable weatherman" would jokingly refer to Wyoming as "that square state." His "fearless forecast" was so popular that WDEF turned their newscast upside down in 1964, so the weather report was at 6:00 p.m., followed by the local news. (WDEF.)

In the 1970s, Channel 12's staff was dominated by veterans who had been with the station since the radio days of the 1950s. The addition of Neal Kassebaum, who was in his early 20s, was a breath of fresh air. His exuberance added a light touch to commercials and the late newscasts, where he handled weather duties with the same irreverent style Harve Bradley had perfected years earlier. (WDEF.)

When Gary Wordlaw was a teenager in 1968, Channel 9 hired him as a stage manager. According to him, he spent a few years "sweeping up after Bob Brandy's horse so we could get the news on." He soon worked his way up to reporter before becoming an assistant news director during his 11-year stay. He has since been a television executive and is currently general manager of WTXL-TV in Tallahassee, Florida. (WTVC.)

By the late 1960s, Channel 3 was well established as the city's top news station, and it ambitiously aired live telethons, parades, and Junior Miss pageants with its remote equipment. In promotional ads, the station touted its veteran news team and the WRCB news cars. Above from left to right are Roy Morris, Don Fischer, Russell Brown, John Gray, David Carlock, Tommy Eason, and Mort Lloyd. (Dorris Prevou.)

Mort Lloyd had been atop the news ratings at WRCB for 12 years when WDEF general manager William Evans gave him something to consider. He offered the anchorman a chance to return to Channel 12, promising more money and news-gathering equipment. On March 16, 1970, Lloyd made the switch to Channel 12, and his viewers followed. Lloyd's new *TV 12 Tri-State Report* immediately became number one. (Morty Lloyd.)

If this 1970 promotional photograph looked familiar to Chattanooga news viewers, it has accomplished its purpose. After returning to WDEF Channel 12, Mort Lloyd (far right) took the same approach he had used at Channel 3, reminding the audience that he was prepared to gather the news by land or by air. Lloyd's WDEF news team colleagues wore red blazers and were dubbed the "Professionals." (WDEF.)

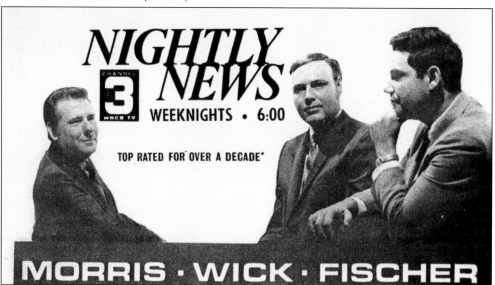

Mort Lloyd's sudden switch from WRCB to WDEF caught his former station by surprise. Two weeks later, weatherman John Gray also made the move. WRCB hurriedly assembled a team which consisted of Roy Morris reporting the news, weatherman Don Wick (who Gray had replaced at WDEF), and Don Fischer remaining as sportscaster. This ad described Channel 3 news as "top rated for over a decade," but that was with Lloyd. (Author's collection.)

The movie *Anchorman* offered a comedic view of a 1970s female reporter invading an all-male newsroom, but Jackie Schulten lived out that scenario in real life. She came to WTVC as a reporter in 1973. Two years later, WRCB offered her a 6:00 p.m. coanchor position. Jackie stayed until 1977, when she decided to attend law school. She is now a Hamilton County Circuit Court judge. (WTVC.)

In March 1973, much of Chattanooga was underwater after 10 inches of rain. WTVC news director Gil Norwood hosted a live special with Mayor Robert Kirk Walker, fielding viewer phone calls and answering questions about the city's flood relief efforts. (WTVC.)

Allen Jones (left) grew up in the Sale Creek area of Hamilton County, and began working at Channel 12 in 1967, anchoring the city's only noon newscast. Channel 3 hired him in 1970 to help fill the void left by the departure of Mort Lloyd. By 1974, he was anchoring the evening news alongside weatherman Don Wick. (Allen Jones.)

The Mort Lloyd–John Gray pairing was so dominant for Channel 12 that the station made a bold move in 1972. A 7:00 p.m. newscast was added, anchored by other members of the news team; however, neither Lloyd nor Gray appeared on the program, and ratings were flat. It soon disappeared, making way for *Dragnet* reruns. (WDEF.)

Despite his 20-year record of topping the local ratings, Mort Lloyd had grown restless by 1974. At the age of 43, he took on a new challenge: a race for Congress. Lloyd was dissatisfied with the Republican Third District incumbent, so he entered the Democratic primary. He garnered 60 percent of the vote against two opponents and began preparing for the general election. (WDEF.)

Three months before the 1974 general election, Mort Lloyd campaigned with son Morty, wife Marilyn, and daughter Mari for the Third District congressional seat. Tragically, Lloyd was killed in a plane crash near Manchester, Tennessee, on August 20. The light plane he was piloting lost a propeller blade, shaking the engine out of its mount. Lloyd was the only person aboard. Marilyn Lloyd, a radio station owner, replaced her late husband on the ballot. She won in November and would serve in Congress for 20 years, never losing an election. (Morty Lloyd.)

In the post–Mort Lloyd era, Channel 12 tapped reporter Mike King (second from left) as its new anchor. King had been news director for WFLI radio in the 1960s. When Lloyd's widow, Marilyn, took up his congressional campaign and won the seat, King joined her office. Seen here with King are, from left to right, Bill Smith, Sharon Summers, and John Gray. (WDEF.)

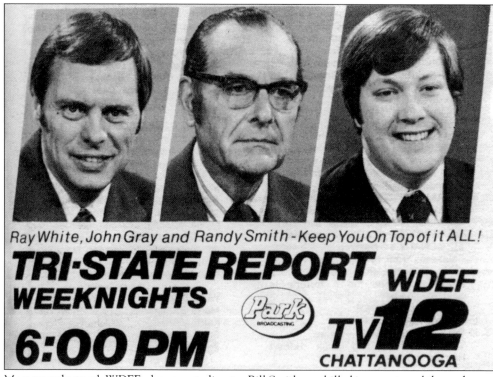

More tragedy struck WDEF when sports director Bill Smith was killed in an automobile crash near Cleveland, Tennessee, in 1975. The new-look team for Channel 12 included Chattanooga news veterans Ray White and John Gray and newcomer Randy Smith, who had been a sportscaster for radio stations WDXB and WDOD. (Author's collection.)

Looking back on this 1972 photograph, Calvin Sneed sees "a skinny 18-year-old kid from Kingsport, Tennessee," who had never worked in television until he was hired by Channel 9. For three years, he reported stories and shot and developed film for the station newscasts. He left the station in 1975 to further his education but returned 17 years later. (WTVC.)

Channel 3's Jerry Wilson got a tip that Muhammad Ali was in Chattanooga in 1975 and snagged an exclusive interview with the boxing champion at the Lakeshore Restaurant. In 1971, Wilson joined the station from WLAR radio in Athens, Tennessee. He was named sports director in 1972 and remained at Channel 3 until 1978. After a stint at WDEF radio, Wilson owned a successful photography business. (Jerry Wilson.)

After 17 years in the ratings cellar, Channel 9 assembled a news team that would catch on with viewers. In 1975, new general manager Jane Dowden hired (from left to right) Don Welch to do the weather and Bob Johnson to anchor the news and promoted Darrell Patterson, a native of Athens, Tennessee, to sports director. Patterson came to Channel 9 in 1974 from WDOD radio. (WTVC.)

Bob Johnson applied and was rejected for jobs at Channels 3 and 12 before Channel 9 hired him in 1975. The 28-year-old anchorman was a teen deejay in Marietta, Georgia, before moving into television at WAGA in Atlanta. There he hosted AM *Atlanta* and the children's show Mr. *Pix*. After several years of musical chairs at the Channel 9 anchor desk, Johnson proved to be a keeper. (WTVC.)

Hollywood lore is full of stories about stars that were discovered while working at a drugstore or waiting tables. Chattanooga's real-life find was Debbie Baer. The bubbly 19-year-old Tennessee Temple college student was a cashier at Red Food Store on Highway 58 when a Channel 12 executive told her she had the personality for television. She landed a weekend weather job and became a local sensation. (WDEF.)

Dayton, Tennessee, native Don Welch was a deejay for WDXB and a news and weather reporter for Channel 3 before his move to Channel 9 in 1975. A perfect fit with Bob Johnson and Darrell Patterson, Welch used his outgoing personality to spice up his weather segments. Shunning the technical bells and whistles, Welch quoted weather wisdom from his "Grandpappy," and trusted the woolly worms to predict when winter would arrive. (WTVC.)

Middle Tennessee native Randy Smith was the Channel 12 sports director from 1978 to 1987. He took a break to become a teacher, then became Channel 3's sports director from 1995 to 2009. He also hosted the University of Tennessee Volunteers radio pre-game and post-game football shows for 17 years and has done play-by-play for hundreds of college football and basketball games for cable networks. (Carson Malone.)

NBC *Today Show* weatherman Willard Scott (center) has made several visits to Chattanooga. His first was in 1983, when he clowned around on the news set with Channel 3's Fred Johnson (left) and Craig Edwards. (Dorris Prevou.)

Channel 9 lengthened its evening newscast for a few years in the early 1980s. The *Action News Hour* ran from 6:00 to 7:00 p.m. and featured, above from left to right, sportscaster Darrell Patterson, news anchors Bob Johnson and Suzy Rigsby, and meteorologist Neal Pascal. Rigsby was Chattanooga's first female radio news director at WGOW in 1975. (WTVC.)

Neal Pascal earned his meteorology degree from Texas A&M in 1977 before coming to Channel 9 from El Paso, Texas, in 1981. For 25 years, he served as Channel 9's chief meteorologist, creating a network of 37 "weather watchers" throughout the station's viewing area. In 2008, he joined Channel 3 as weekend meteorologist, leaving in 2010 for a teaching career. (WTVC.)

From 1978 to 1981, North Carolinian Sky Yancey provided a stable presence at the anchor desk during a tumultuous time at Channel 3. She told a newspaper reporter that working for four news directors and with three male coanchors had been a challenging experience. She later anchored the news for a station in Lexington, Kentucky. She got the name "Sky" from her initials: Sarah Kendall Yancey. (Dorris Prevou.)

Award-winning South Dakota broadcaster Dale Harris came to Channel 12 as an evening news anchor in 1979 but was soon given additional responsibilities as news director. Harris kept the longtime news leader in first place until WTVC surged ahead in 1982. Harris left the station in 1983. (WDEF.)

Channel 9's Bob Johnson spent seven years trying to get the newscast out of the ratings cellar. His work paid off, and by 1982, *Action News* was in first place. In February 1983, Channel 9 pulled a major coup, hiring popular anchor Tracy Moore away from Channel 12. The Alaska native stayed only two years before taking a job in Tampa, Florida. She is now a family court judge in Tampa. (WTVC.)

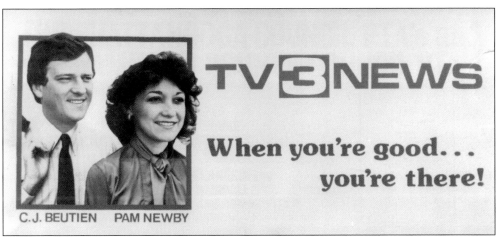

From 1981 to 1983, Channel 3 went with the youthful anchor team of C. J. Beutien and Pam Newby. Beutien was from the Chicago area and would go on to become Channel 3's assignment editor and a successful news director in Ohio and Indiana. Newby was a radio newscaster for WFLI before becoming a reporter for Channel 3. (Dorris Prevou.)

Channel 3 changed its strategy in 1983, going with an all-male anchor team. From left to right are Bill Markham, Fred Johnson, Jim Thrasher, and Tony Shelton. Markham had recently worked for WHNT-TV in Huntsville, Alabama. Johnson, a local product, is a highly decorated 14-year Marine Corps veteran. Thrasher is also a Chattanooga native. Meteorologist Shelton replaced a series of weather girls. (WRCB.)

Much like Channel 3's earlier commitment to live, on-scene programming, the station was the first to invest in an ENG (electronic news gathering) truck. In the early 1980s, WRCB used its bright orange "live eye" to broadcast everything from department store grand openings to icy road conditions. The station promoted its live capabilities relentlessly, boosting its image as the channel to watch for breaking news. (Dorris Prevou.)

By 1983, Channel 3 had gone through 10 male news anchors in as many years. There was also heavy turnover among its news directors, weather forecasters, and sportscasters, so the station was seeking stability. Alabama native Bill Markham answered the call, beginning a 26-year run highlighted by his weekly "Crimestoppers" segment, which in his words, "put the bad guys behind bars." (WRCB.)

When Channel 3 management suggested a female coanchor for Markham in 1985, he called on an old friend. Cindy Sexton had worked with Markham in Huntsville, where they had been a winning team. She had moved on to Las Vegas and was ready to return to the South. The Murray (Kentucky) State University alumna made her Chattanooga television debut on *Eyewitness News* in September 1985. (WRCB.)

Paul Barys graduated from Northern Illinois University with a meteorology degree in 1973. After working for television stations in North Carolina and Indiana, he moved Cleveland, Ohio, and began working at WKYC-TV alongside future NBC weatherman Al Roker. In 1985, Barys became chief meteorologist at Channel 3, with the popular catchphrase, "Paul said it would be like this." His "Snowbird Report" was required viewing for children seeking a day off from school. (WRCB.)

Like many anchors, Jerry Brown got his start in radio as a teenager at his hometown station in Rome, Georgia. After serving as a military journalist in the U.S. Army Reserve, Brown anchored in several cities before landing at Channel 12 in 1985. He left after three years but returned for a second stint from 1998 to 2003. He is now the evening news anchor at WMBB in Panama City, Florida. (WDEF.)

David Neal had just graduated from Florida State University in 1987 when Channel 12 recruited him for the chief meteorologist job. Since John Gray retired in 1977, the weather position had been a revolving door. Neal, from Gadsden, Alabama, enjoyed being close to home and stayed at Channel 12 for 10 years. Since 1997, he has worked for television stations in Birmingham, Alabama. (WDEF.)

Ray White got his start in Chattanooga and kept coming back. In 1959, the University of Chattanooga student began on WDEF radio, switching to the television side in the 1960s. After serving two terms in the Tennessee Legislature, he returned as Channel 12's news director and anchor from 1975 to 1979. He held both jobs again from 1985 to 1988 and then finished his career as news anchor in Mobile, Alabama. (WDEF.)

WRCB continued setting the pace with live broadcasting equipment by purchasing the market's first satellite truck in 1988. With satellite commander Bobby Winders at the controls, Channel 3 began airing live broadcasts from far outside the Chattanooga viewing area. Reporters did live shots from the presidential inauguration, the National Aquarium in Baltimore, and tornado-ravaged Huntsville, Alabama. Both competing stations soon got their own satellite trucks. (WRCB.)

Channel 3 scored another first in 1987, adding a half-hour local newscast prior to the traditional 6:00 p.m. start time. *Live at 5:30* offered a quick look at the news, along with features on health, cooking, and entertainment. Anchors included David Carroll, who had been with Channel 12, and Tamara Lister, the station's consumer reporter. Not to be outdone, Channel 9 began its own 5:30 newscast one week later. (WRCB.)

In 1985, Channel 9 again hired a popular competitor for its female evening news anchor position. Pam Newby came over from Channel 3, anchoring with, from left to right, Neal Pascal, Bob Johnson, and Darrell Patterson. Newby was with the station until 1988. She later reported for the Christian Broadcasting Network. (WTVC.)

Channel 3's Frenche Brewer was hired as a reporter in 1985 and was soon promoted to weekend news anchor. By the time she left the station in 1993, she had earned a reputation as a solid journalist and a stylish dresser as well. Today Brewer is in her home state, working in media relations at the University of South Carolina. (WRCB.)

Cater Lee coanchored the evening news on Channel 12 from 1986 to 1988. She left Chattanooga for California, where she has anchored and reported for CBS-2 and KCAL-9 News in Los Angeles. She has also produced long-form pieces for NBC's *Today Show* and is a five-time Emmy Award winner. (WDEF.)

Channel 3's dynamic sports team of the late 1980s featured Bryan Houston (left) and Russ Riesinger. Houston is a Louisiana native who worked in Texas television news before and after his Chattanooga tenure. Riesinger, originally from Montana, earned his degree from the University of Tennessee at Chattanooga while at Channel 3. He has anchored the news at WSAV in Savannah, Georgia, since 2004. (WRCB.)

Bill Markham and Cindy Sexton took Channel 3's *Eyewitness News* on the road in the summer of 1986. "Live From Your Hometown" featured the news team at courthouse squares throughout the Tennessee Valley. Each newscast featured the local flavor of towns ranging from Dayton, Tennessee, to Ringgold, Georgia. Both anchors said the best part of the promotion was meeting their viewers in person. (WRCB.)

In 1983, David Glenn (left), then a 16-year-old Gordon Lee High School student from Chickamauga, Georgia, visited Channel 9. Glenn wanted to be a weatherman, so he asked chief meteorologist Neal Pascal for advice. It worked, as Glenn landed his first job at Channel 3 in 1990, and after moving to Mobile, Alabama, he returned to Chattanooga in 2008 as chief meteorologist at WTVC. He still keeps this photograph on his desk. (David Glenn.)

Channel 9 had a solid lineup of weekend anchors in the 1980s. Seated is John Gilbert, and standing from left to right are John Favole, Dave Staley, and meteorologist George Matthews. Sportscaster Staley has won multiple Edward R. Murrow and Emmy awards for his popular "Dave's Diamond Darlings" feature, which spotlights children learning to play baseball. Staley and Darrell Patterson make up the area's longest-running television sports team. (WTVC.)

Channel 12 endured another anchor team turnover in 1988 but came up with some talented replacements. Above from left to right are meteorologist David Neal, news anchors Christy Murphy and Bill Ross, and sports director Kevin Billingsley. Murphy would stay until 1992, while several male coanchors would come and go. Billingsley, from Whitwell, Tennessee, would remain at WDEF for 10 years. (WDEF.)

WRCB and WTVC stayed neck-and-neck in the ratings race of the 1990s, with each team experiencing very little turnover. Channel 3's team in 1991 included, from left to right, Paul Barys, Cindy Sexton, Bill Markham, and sports director John Fricke. Fricke, a veteran of CNN, stayed at Channel 3 until 1995. In recent years, he has hosted national talk shows on Fox Sports Radio. (WRCB.)

A fast-paced television newscast requires a steady hand at the controls, and Wayne Jackson has certainly been a stable fixture at Channel 3. Jackson has directed the evening news at Channel 3 since 1974. This feat is even more amazing when one considers the dazzling advances in technology since that time. Jackson has changed with the times, making Channel 3's news broadcasts look cutting-edge each evening. (Dorris Prevou.)

As television news departments added more programs, the stations began employing three-person weather teams in the 1990s. Channel 12's forecasters at the time were, above from left to right, Chip Chapman, David Neal, and Patrick Core. Chapman, a Chattanooga native, had worked in local radio at KZ-106 and US-101. The Louisiana-born Core came to Channel 12 as a weekend weatherman in 1990 and has been the station's chief meteorologist since 1997. (WDEF.)

Channel 3's *Eyewitness News Today* team in the early 1990s included, from left to right, anchor Jed Mescon and weather forecasters Mary Kate Wells and David Glenn. Wells, who was also a reporter, was pressed into duty as a third weather person when the station expanded its morning and weekend newscasts. In 1994, she got married, moved to Connecticut, and left the television business. (WRCB.)

From 1992 to 1995, Channel 12 countered their competitors' long-running anchor teams with the duo of Garry Mac and Rebecca Williams. Mac was a well-known local radio personality, dating back to his days on WDXB in the 1970s. Williams had worked her way up the ladder at Channel 12, starting as a studio camera operator on the *Morning Show* in 1987 before becoming a reporter. (WDEF.)

Channel 12 went with a combination of experience and youth behind its anchor desk from 1996 to 1998. Tennessee native Bill Mitchell (left) was brought in to team with 27-year-old Heidi Robinson, who had been a consumer reporter at Channel 3. Mitchell remains with Channel 12 today as a reporter. Meteorologist Patrick Core is on the right. (WDEF.)

In the late 1990s, third-place WDEF retooled again. From left to right are Patrick Core, Sally Schulze, Glenn Halbrooks, and sportscaster John Appicello. Schulze came to Channel 12 from Illinois and later anchored in Orlando before leaving television. Halbrooks, a McCallie school alumnus, is now a news anchor at WAKA-TV in Montgomery, Alabama. Appicello, from Pennsylvania, stayed at WDEF for five years and is now in Roanoke, Virginia. (WDEF.)

Fox affiliate WDSI-61 tried its hand at local news in 1999, breaking away from a partnership with WTVC to produce its own morning, midday, and nightly newscasts. A 30-person team was assembled, giving Chattanooga four television news departments. By 2004, the experiment ended, and the station reteamed with WTVC, using that station's staff for a 10:00 p.m. newscast. Anchor Dan Howell (front) now hosts a weekend talk show on WDSI. (Dan Howell.)

Seven

You're Like Family to Us

"We wouldn't start our day without you."
"I grew up listening to you on the radio."
"You all are like part of our family."

Broadcasters never tire of hearing compliments like these. Such accolades reinforce their reasons for getting into the communications field. The most successful television and radio personalities are those who are able to see through the camera or break the barrier between the studio microphone and the listener.

By its very nature, broadcasting is designed for people on the move. As the preceding pages illustrate, Chattanooga is often a training ground for journalists and entertainers who are working their way up to a large market station or even a network job; however, the city is also blessed to have many media personalities who have chosen to stay, whether it be for family reasons, professional satisfaction, or just a love of the area.

These enduring faces and voices are among those who have made a personal connection with Chattanooga viewers and listeners. Their durability is proof of the power of personality.

Channel 3's Jed Mescon (right) is shown with the station's first employee, Wayne Abercrombie, at WRCB's 50th anniversary luncheon in 2006. Mescon left his hometown of Atlanta for Arizona State University. After graduation, he worked in Savannah, Georgia, and Charleston, South Carolina, before settling in at WRCB in 1987. His enthusiastic style provides a daily jolt of energy for morning viewers of *Eyewitness News Today*. (Jed Mescon.)

In 1995, Channel 3 was seeking a nighttime reporter and hired Nashville native LaTrice Currie who was working at a station in Jackson, Tennessee. She was soon promoted to weekend anchor and became coanchor of *Eyewitness News Today* in 1999. She also presents nightly health reports on *Live at 5*. (Rick Owens/WRCB.)

Channel 9's Darrell Patterson has been a Chattanooga favorite since joining the station in 1974. Long known for presenting his sportscast while standing at a podium, Patterson's commanding voice projects a genuine love for the games he covers. His annual Volunteer Challenge Golf Tournament has raised much-needed funds for Santa For All Seasons since 1990. He is proud to call longtime colleague Bob Johnson his best friend. (WTVC.)

For more than 25 years, Channel 9 anchorman Bob Johnson introduced viewers to children in need of a Big Brother or Big Sister in his *Wednesday's Child* segment. The stories brought out the kid in the newsman, who participated in games, rides, and sports with his young friends. After being diagnosed with Parkinson's Disease, Johnson retired in 2007, but he is still greeted warmly by viewers everywhere he goes. (WTVC.)

Radio has dominated Patti Sanders's life for 30 years, and she wouldn't have it any other way. During most of that time, she's been the bright midday voice on light rock stations that are heard in almost every doctor's office in town. Since the early 1990s, she has spread good cheer on Sunny 92.3 (WDEF-FM). If she has ever had a bad day, her listeners were unaware of it. (Patti Sanders.)

"Dr. John" Baker played the soul hits on WNOO for 25 years and is now on Groove 93 WMPZ-FM. Married with four children and seven grandchildren, Baker says radio is his true calling, and he plans to stay on the air "until I'm too old to walk in the studio."

In 1983, WUSY (US-101) made its debut, and Chattanooga took immediate notice. The station has topped the ratings for the past 20 years. In 1998, the announcers celebrated a fourth consecutive CMA Station of the Year award. Clockwise from left are Ken Hicks, Brandy, Bearman, Dex, Tag Martin, Erin Michaels, and in the center, the late David Earl Hughes. Hicks, Bearman, and Dex are still with the station today. (WUSY.)

Since 1996, Bearman (left) and Ken Hicks (right), seen above with singer Brad Paisley, have been a top-rated morning team for US-101. Bearman was a midday deejay for the station before moving to the morning show, and Hicks was the host of the station's Sunday *Gospel Road* program, which he still hosts to this day. Hicks also sings and emcees gospel music shows throughout the tri-state area. (WUSY.)

Chattanooga's senior news anchor Cindy Sexton consistently wins reader polls for Best Television Personality. In 2008, she added another popular segment to Channel 3's *Eyewitness News at 6.* In "Forever Family," she helps adoptable children from all over Tennessee find loving homes. (WRCB.)

These two WFLI guys from 1961 are still best friends today. Tommy Jett (left) and Johnny Eagle meet up regularly at radio reunions to talk about the days when stations were locally owned and doing outlandish promotions. Jett still produces an oldies show that can be heard on his Web site, www.tommyjett.com. (Tommy Jett.)

Since coming to Chattanooga in 1985, Channel 3 chief meteorologist Paul Barys has welcomed more than 1,200 children into the studio as junior forecasters, helped give away more than a thousand dogs with his "Paul's Pet" segment, and served up countless plates at his Backyard Barbecues. These activities, along with his pinpoint forecasts, have made him the city's favorite forecaster according to numerous reader polls. (WRCB.)

From his teenage days in the 1950s to his retirement in 2009, no one in Chattanooga witnessed more first-hand history than Tommy Eason. The fearless news photographer shot civil rights marches, the 1964 Jimmy Hoffa trial, famous figures, infamous criminals, and every weather event that invaded the Tennessee Valley. Eason keeps busy in retirement as a freelance photographer and a devoted family man. (Amy Morrow.)

The once-turbulent Channel 12 anchor desk settled down in the 2000s. From left to right are weatherman Patrick Core, anchors Candice Lee and John Mercer, and sports director Rick Nyman. Core has been a steady presence since 1990. Lee came to WDEF from the U.S. Navy and stayed from 1997 until 2007. Mercer, who hails from Washington, and Nyman, an Alabama native, both joined the station in 2003. (WDEF.)

WTVC's longtime morning news team split up in the early 2000s but are still visible on local screens. Don Welch (left), having weathered many fashion and hairstyle changes over the years, is host of the midday *This N That* show on Channel 9, while Bill Race is the station's senior meteorologist, seen each morning on *Good Morning Chattanooga*. Melydia Clewell left Channel 9 in 2009, joining Channel 3 as an assignment manager and reporter. (WTVC.)

MaryEllen Locher (right, with Kathie Lee Gifford) was a proud Penn State alumna who came to Channel 9 as a health reporter in 1985. By 1991, she was coanchoring the evening news. Her male coanchors called her "Mel," quickly accepting her as "one of the guys," according to colleague Bob Johnson.

In 1988, Locher was diagnosed with cancer and courageously underwent treatment for the next 17 years. On the nightly newscasts, she shared details of her cancer battle with the viewers. She retired from WTVC on June 7, 2005, her 20th anniversary with the station. Two days later, she died at the age of 45. The MaryEllen Locher Breast Center at Memorial Hospital is a fitting tribute to an inspirational broadcaster. (WTVC.)

News anchor Mike Dunne (left) and sportscaster Dave Staley were longtime colleagues on WTVC's weekend news. Dunne joined Hamilton County government in 2006 after working as a weekend anchor for 17 years. Staley added news anchoring to his duties after Dunne's departure and can still be seen on the station's weekend news today. (WTVC.)

Calvin Sneed started his second stint at WTVC in 1992 as the station's consumer reporter. He anchored the 5:30 p.m. newscast until moving to the 6:00 p.m. slot in 2007, when Bob Johnson retired. One of Sneed's popular features was "Does It Work?" In this segment, Sneed would try out products to see if they performed as advertised. (WTVC.)

WRCB's Bill Markham (right) anchored *Eyewitness News at 6 and 11* from 1983 to 2006 and *Live at 5:30* from 2006 until his retirement in 2009. He is shown here with Pat Boone at the singer's annual golf tournament for Bethel Bible Village. Markham participated in the Boone fund-raiser every year. (Bill Markham.)

One of the perks of working in local television is meeting celebrities who travel around the country to promote charitable causes. WRCB's *Eyewitness News at 6* anchors David Carroll and Cindy Sexton had the privilege of hosting fitness guru Richard Simmons in 2006. (Author's collection.)

"Magic" Crutcher (left, with WJTT owner Jim Brewer) is one of the most familiar voices in Chattanooga radio, and is still at WJTT Power-94 after almost 30 years. The Chattanooga native is the smooth hostess of the midday show, as well as the music director for the station. (WJTT.)

One of the secrets to the success of WJTT is its personalities' strong connection to the community. Eric Foster is another local product who has entertained the station's listeners since 1994. Now hosting the *Chattanooga Morning Show* with Donna L., Eric is also a busy musician in his own right. (WJTT.)

Kim Chapman first came to Chattanooga television in 1994, coanchoring a WTVC-produced 10:00 p.m. newscast on Fox affiliate WDSI. She moved to Jacksonville, Florida, in 1997 and then returned to work for Channel 9 in 2000. She frequently substituted for MaryEllen Locher during her illness and succeeded the beloved news anchor when she died in 2005. Chapman is an active volunteer for the Epilepsy Foundation. (WTVC.)

Rossville, Georgia, native Amy Morrow began her news career right out of the University of Georgia. She worked her way up from stations in Dalton, Georgia, and Tupelo, Mississippi, to Chattanooga's WDSI in 1999. Two years later, she joined Channel 3. She is now an anchor and the station's "Crimestoppers" reporter. (Rick Owens/WRCB.)

Two of Chattanooga's broadcasting greats, who often competed with each other on radio and in television in the 1950s and 1960s, are in this 2004 photograph. Luther Masingill (left) and Roy Morris achieved success in both mediums, each with a long, distinguished career. Morris died in 2006 at the age of 85. (Earl Freudenberg.)

Luther Masingill (left) was at the microphone during the attack on Pearl Harbor in 1941, as well as in 2001 during the attack on the World Trade Center. His longevity is unequaled. Fellow Chattanoogan James Howard joined Luther's *Sundial* program in 1993. Howard has made several holiday season trips to Iraq, taking care packages to Chattanooga-area servicemen and women. (Tom Smith/WDEF.)

After appearing on Chattanooga television with his trademark mustache for more than 25 years, Channel 9 anchor Bob Johnson (third from left) surprised coworkers and viewers by showing up for work clean shaven. The next day, his colleagues Neal Pascal, MaryEllen Locher, and Darrell Patterson had a little fun with Johnson by adopting his old look. (WTVC.)

More than three decades after her *Funtime* show left the airwaves, Channel 9's Marcia Kling is still a favorite of local viewers. Many remember her courageous recovery from a rare form of mouth cancer in 1974, when she had to relearn to speak. Others thank her for teaching them to read or how to do arithmetic. And still others consider her a wonderful role model and mentor. She was named Tennessee's Woman of Distinction in 2008 and is still a daily presence on Channel 9 today. (WTVC.)

Discover Thousands of Local History Books
Featuring Millions of Vintage Images

Arcadia Publishing, the leading local history publisher in the United States, is committed to making history accessible and meaningful through publishing books that celebrate and preserve the heritage of America's people and places.

Find more books like this at
www.arcadiapublishing.com

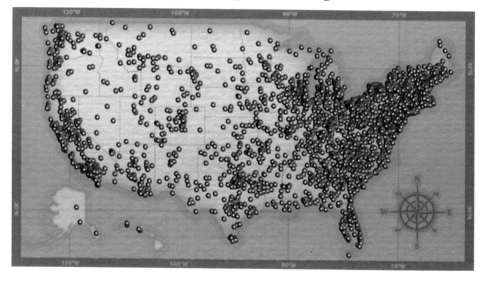

Search for your hometown history, your old stomping grounds, and even your favorite sports team.

Consistent with our mission to preserve history on a local level, this book was printed in South Carolina on American-made paper and manufactured entirely in the United States. Products carrying the accredited Forest Stewardship Council (FSC) label are printed on 100 percent FSC-certified paper.

MADE IN THE USA